Living with the
weather

Snow and Ice

Philip Steele

WAYLAND

Living With the Weather
SNOW AND ICE

Other titles:
FOG, MIST AND SMOG
RAIN, WIND AND STORM
HEAT AND DROUGHT

Produced for Wayland (Publishers) Ltd
by Thunderbolt Partnership
6 Blackstock Mews, London N4 2BT

Designed by Flora Awolaja
Edited by Belinda Weber
Illustrated by Sebastian Quigley
Sue Sharples, and Mary Hall
Picture Research by Veneta Bullen

British Library Cataloguing in Publication Data
Steele, Philip
   Snow and Ice. - (Living With The Weather)
   1 Snow - Juvenile literature 2. Ice - Juvenile
   literature
   1. Title
   551.5'784
ISBN 0 7502 2021 X

Printed and bound in Italy by EUROGRAFICA S.P.A.,
C.S.p.A, Italy.

CONTENTS

WHITEOUT

'Storm!...It crashes, sprays, howls. In the morning it is nearly impossible to open the tent flap, the snow-drift has it 'walled up'. Outside, whiteout. Not mist, but snow plumes so thick that we can see only a few metres...'
This was one diary entry written by ace explorer Reinhold Messner, as he walked across the icy wilderness of Antarctica in January 1990.

Studying the cold
Over the ages, people have learned to survive in the ice and snow. Some people make their homes in the Arctic, but nobody lives in the bitter lands of Antarctica, except for scientists who visit the bases there. They look at the way the polar regions affect the weather in the rest of the world. Ice and snow cover nearly a quarter of our planet, either all the time or just in winter.

The Antarctic is the coldest place on ▲ Earth. Buildings often freeze completely.

► Sometimes, when the clouds are low, it is impossible to see where the clouds end and the land begins. This is called whiteout.

WHAT IS ICE?

The big freeze

To measure cold and warmth, we use thermometers. These are usually glass tubes, with a bulb at the bottom which is filled with a liquid metal called mercury. As the metal becomes warmer, it swells or expands. The silvery metal rises up the tube. As it cools, the mercury shrinks or contracts and falls down the tube.

Measuring by degrees

The tube is marked in units of temperature called degrees (°). The most commonly used degree system is called Celsius (C) after the Swedish scientist Anders Celsius who invented it in 1742. At 100°C, water boils. At 0°C, it freezes. In nature, water is usually in a liquid state, able to flow freely. It may also hang in the air as vapour. But when it freezes, it becomes solid and is called ice.

ICE AND WATER

1 Half fill an ice tray with water. Mark the level on the outside of the ice tray. Place it in the freezer box of a fridge.

2 When it has frozen, does the ice take up more or less space than the water?

3 The ice is less dense than water. Put the ice cubes in a bowl of water. Do they float or sink?

▲ **In some countries, including the Netherlands, canals, rivers and lakes may freeze in the winter. If the ice is thick enough, people can skate on the surface.**

Frozen seas

Ice is cold and clear, hard, shiny and slippery. When the temperature drops below 0°C, ponds and lakes freeze over and dripping water turns into icicles. The salt in seawater makes it harder to freeze, and so oceans do not begin to ice up until the water reaches a temperature of –2°C. Oceans take a long time to cool down, so it is only in polar regions that the sea freezes regularly. When the temperature rises, there is a thaw. The ice melts and turns back to water again.

Hidden ice

★ Great blocks of ice called icebergs float through the polar seas. The biggest ones can be over 8,000 square kilometres in area, the size of a small country. Up to 90 per cent of the iceberg is often under the water.

★ In places the ice covering the coast of Antarctica is over 4.5 kilometres thick. And that's about the height of 3,000 children!

▲ Huge chunks of ice sometimes break off from glaciers and float out to sea. They can cause problems for shipping as most of their bulk is hidden under the water. Ships use sonar to tell them whereabouts the icebergs are to be found.

ICY WEATHER
Crystals of ice

Low temperatures can bring about all sorts of different weather conditions. The dew may freeze on the ground, covering grass and plants with a glittering coat of ice crystals called frost. Fog and rain may freeze on the twigs and branches of trees, making them white.

Snowflake stars

Snow is made when ice crystals form in clouds. The ice makes tiny shapes such as needles, blocks and rods. It takes millions of these to build up a beautiful star-like structure, often with six sides or points. Each one of these snowflakes is different. Snowflakes often stick together as they fall to the ground. There they may build up into a thick, white layer. There are lots of different kinds of snow. It can be wet and mushy, crunchy, or light and powdery. Powdery snow is the favourite for skiers!

◄ **Thick snow can take quite a while to melt as its white colour reflects most of the sunlight away. When the snow is frozen and crunchy, sledging is a fun pastime.**

Blizzards and snow drifts

High winds may drive the snow at an angle, creating a blizzard. It is hard to walk or drive, or even to see more than a few metres ahead of you. The wind may whip up snow lying on the ground, piling it into deep drifts along the edges of roads or by the side of fields. Sleet (a mixture of snow, ice and rain) may pour down as the temperature falls. Sleet is wet and splodgy. Hail is much harder than sleet and snow. The hard, white icy pellets can sting your face or rattle against the windows.

The temperature rises

When the temperature rises and the thaw comes, snow and ice turn into slush and melt. The water may flow into rivers and lakes and sometimes causes flooding. In cities, drains may overflow and streets may flood.

▲ **Hailstones form within clouds such as this one. Small hailstones are usually round and about the size of a pea, but larger ones can be almost any shape.**

Safety first!

★ Icy weather is fun – but it can be dangerous! Humans need warmth to survive. Always wrap up well and wear waterproof clothing.

★ Wear strong boots which will grip the snow and ice. One slip and you could break a leg or an arm.

★ If you start feeling too cold, go back indoors and have a warm drink.

★ Never walk out on to frozen ponds unless the ice has been passed as safe for skating. The ice may break under your weight and trap you under the water.

★ In 1911 the depth of snow lying on the ground at one spot in the American state of California was measured at nearly 10.5 metres. That's about the height of seven children...

WHERE AND WHY?

Cold tops and bottoms

The world's coldest regions lie around the poles, which are always covered with ice-caps. The South Pole lies on the continent of Antarctica, whose ice-cap contains ninety per cent of all the world's ice. Antarctica is completely surrounded by oceans. The North Pole lies on the permanently frozen Arctic Ocean, which is surrounded by the northernmost parts of Europe, Asia and North America.

These northern regions include vast areas of tundra, treeless plains where the soil is permanently frozen to a depth of up to 1,000 metres. The frozen soil is called permafrost. For most of the year the tundra is covered in ice and snow, but during the brief Arctic summer there is a surface thaw. Melting ice forms pools visited by migrating birds, and wildflowers burst into bloom.

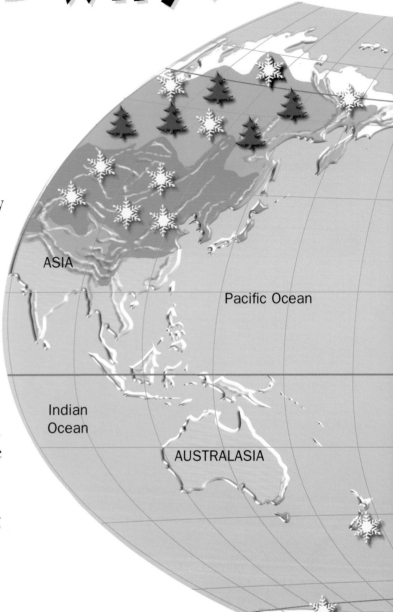

ASIA

Pacific Ocean

Indian Ocean

AUSTRALASIA

Freezing cold winters

To the south of the tundra is a broad belt of forest, stretching through Scandinavia, Russia and North America. Winters here are bitterly cold and snow covers the ground. Southwards again lie lands which have a more temperate or mild climate.

Even though these lands have warm or moist summers, they may still have cold weather in winter, with frost and snow. Winter snows also cover the cold deserts and grasslands which stretch from Central to Eastern Asia.

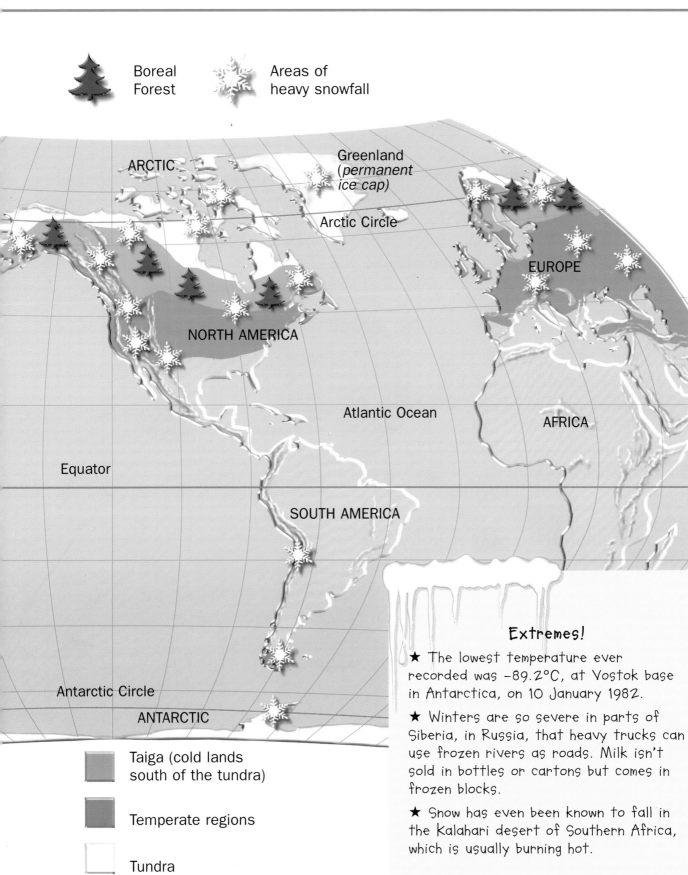

Boreal Forest

Areas of heavy snowfall

ARCTIC

Greenland
(*permanent ice cap*)

Arctic Circle

EUROPE

NORTH AMERICA

Atlantic Ocean

AFRICA

Equator

SOUTH AMERICA

Antarctic Circle

ANTARCTIC

Taiga (cold lands south of the tundra)

Temperate regions

Tundra

Mountainous regions

Extremes!

★ The lowest temperature ever recorded was −89.2°C, at Vostok base in Antarctica, on 10 January 1982.

★ Winters are so severe in parts of Siberia, in Russia, that heavy trucks can use frozen rivers as roads. Milk isn't sold in bottles or cartons but comes in frozen blocks.

★ Snow has even been known to fall in the Kalahari desert of Southern Africa, which is usually burning hot.

LOCAL CONDITIONS
What's the weather like near you?

'Great God! This is an awful place...' wrote the British explorer Robert Falcon Scott, on reaching the South Pole in 1912. He died on the return journey.

How close is the Sun?

Why do some parts of our planet have such a terribly cold climate, while others are burning hot? The climate of a region is the pattern of weather recorded there over a long period. It is affected by many different factors. Some of these are global, while some are local. The closeness of the Sun affects the weather.

As the Earth travels around the Sun, the polar regions are the furthest from its warming rays, so they remain cold. Because the Earth is slightly tilted, some parts pass nearer to the Sun and experience warm, summer months. The parts facing away from the Sun experience the cold of winter.

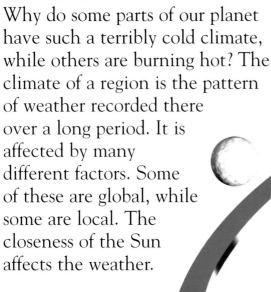

Earth

Moon

Sun

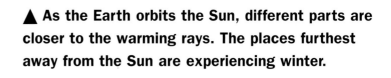

▲ As the Earth orbits the Sun, different parts are closer to the warming rays. The places furthest away from the Sun are experiencing winter.

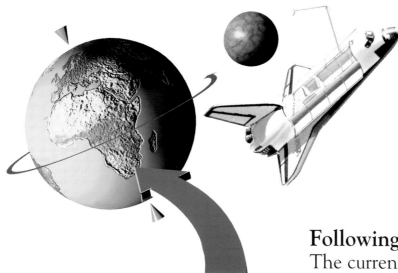

Following the currents

The currents of water flowing through the oceans also affect the weather. Cold water streams from the poles towards the warm currents of the tropics. In the same way currents of cold and warm air flow between the polar regions and tropics, creating the winds.

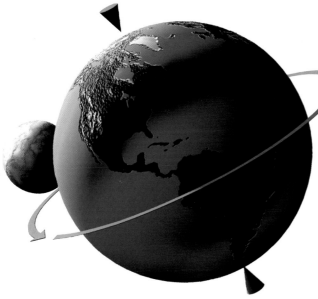

▶ **The warm and cold currents that flow through the world's oceans can affect the temperature of the air.**

 Warm currents   Cold currents

ABOVE THE SNOWLINE

One of the most important factors affecting the local climate is altitude. The higher you climb, the colder it gets. Plateau regions such as Tibet, in the Himalayan mountain range of Asia, experience very low temperatures and heavy falls of snow. Highlands have a cooler climate than lowlands, even in temperate parts of the world. Even when a mountain lies near the Equator, its peaks may remain permanently covered in snow and ice. Mount Kilimanjaro, 5,950 metres above sea level, is snowy all the year round, even though it overlooks the baking hot plains of East Africa.

▲ The tops of mountains, such as Mount Cook and Mount Tasman in New Zealand, are covered with snow all year round.

Deep freezes

★ Altitude can affect city weather, too. Sometimes it can be raining as you enter a skyscraper at street level, but when you leave the lift on the top floor you find that it is snowing outside.

★ In 1991 melting ice in a glacier in the Alps, on the border of Italy and Austria, revealed a corpse. Before a murder enquiry was started, it was discovered that this was the body of a Stone Age man who had been climbing the mountains about 5,000 years ago. His body had been preserved by the cold of the ice.

The higher you climb

As you climb the side of a high mountain, the vegetation changes as the temperature becomes colder. Climbers will notice the same kind of changes that they would see travelling northwards through the lands of the Arctic. Pine forest gives way to the frozen, treeless soil of the Alpine tundra, and this gives way in turn to the bare rock, snow and ice around the summit.

Snow which falls into mountain crevices may become packed down over the ages until it forms a solid mass of ice. As this becomes heavier, it begins to inch its way down the mountainside, forming a river of ice or glacier. Sometimes when the temperature rises heavy falls of snow break loose from the mountainside and tumble down into the valley with a great roar. These are called avalanches.

Lower down the mountain, there may be a temperate forest

The bottom zone on the mountain is deciduous forest with dense plant growth.

▶ **Mountains have broad zones in which different plants and animals live.**

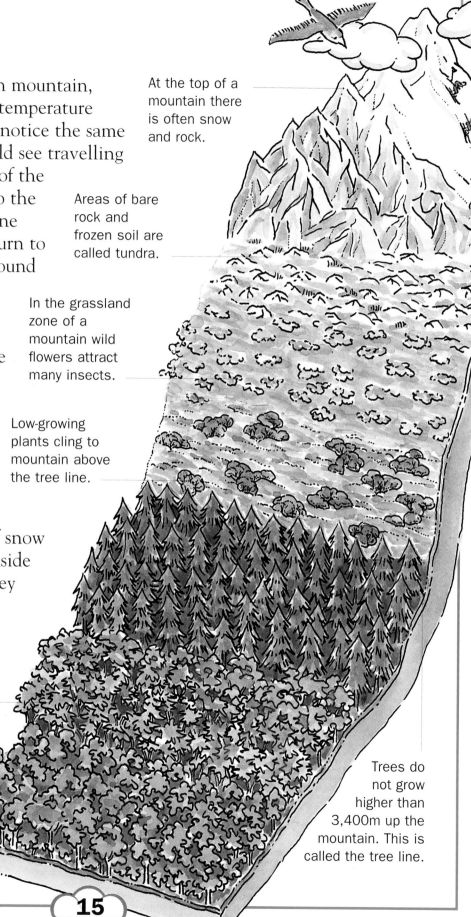

At the top of a mountain there is often snow and rock.

Areas of bare rock and frozen soil are called tundra.

In the grassland zone of a mountain wild flowers attract many insects.

Low-growing plants cling to mountain above the tree line.

Trees do not grow higher than 3,400m up the mountain. This is called the tree line.

SNOW SCIENCE

So just how does cold weather work? What makes it snow? It is all a question of water and temperature. Over seventy per cent of our planet is covered with water, which is stored in the rivers, lakes, seas and oceans. As this is heated by the Sun, some of it turns into vapour. This is called evaporation, which you can see taking place when a puddle of rain dries out in the Sun. The air we breathe is full of water vapour. The water vapour is invisible – until the temperature falls.

Making rain and snow

Warm air rises, carrying water vapour from the sea with it. But the higher the air rises, the cooler it becomes. The vapour condenses, turning back into tiny drops of water. These drops hang in the air as clouds. When they become cooler still, the drops of water will fall as rain or snow – a process called precipitation.

▶ **Every single snowflake is different from all the rest, but they are all six-sided stars. The largest snowflake ever recorded was 38cm across. It fell in Montana, USA, in 1887.**

Cycling water

Freezing temperatures turn the water vapour in clouds into feathery ice crystals, which stick together as flakes and float to the ground as snow. When winds which are heavy with moisture from the ocean meet a mountain barrier, they are forced upwards and cooled rapidly, making snow more likely on mountain slopes. When fallen snow melts, it fills streams and rivers, which flow back to the sea. The water cycle has been completed and can start again.

◄ **Snow can block roads and stop the traffic. This Japanese man is clearing the snow so that he can drive his car.**

When the water vapour condenses, clouds form.

The wind can carry the clouds for great distances before the cloud releases the water as rain.

Water evaporates from lakes, rivers or oceans.

Plants release water vapour through their leaves.

Deep, deep snow
★ The greatest average snowfall ever recorded in one place over a year was 31,102 millimetres. It fell on the slopes of the 4,392 metre high Mount Rainier, in the American state of Washington, in the winter of 1971–72.

Clouds release their water over high ground. It may fall as rain or snow, depending on the land temperature.

The falling water seeps into the ground or collects in rivers, lakes or oceans and the cycle begins again.

FROSTY FACTS

What are hailstones?

Hail is another kind of precipitation. Ice crystals may grow around specks of dust inside a freezing cloud. They chill the droplets of water near them and gather them together, until they make a ball of ice which is heavier and harder than a snowflake. In some tall, billowing clouds, like the ones seen in thundery weather, air currents can shoot upwards at speeds of up to 100 kilometres per hour. The hailstones whizz up and down, gathering more and more ice which builds up like the layers of an onion. Their heavy weight then makes them drop to the ground like bullets.

Icy dew

Sometimes during the night, the temperature of the air drops so low that the water vapour it holds condenses and turns into drops of water. This is known as the dewpoint – the drops of water are called dew.

By the next morning the ground is covered in the droplets of water we call dew. If the temperature of the ground is cooled to a point below 0°C, the dew forms ice crystals, covering the ground and plants in frost.

Upcurrents of air carry the hailstone back to the top of the cloud.

A new layer of ice is added to the hailstone on each trip through the cloud.

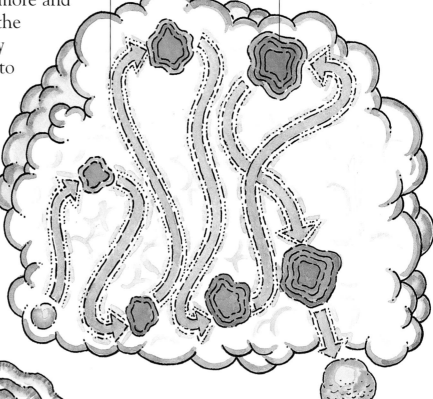

▶ The number of layers in a hailstone shows how many times it has been coated with ice.

▲ Hailstones drop to the ground when they are so heavy that the upcurrents of air in the cloud cannot support them.

▲ On frosty mornings, feathery ice crystals cover every surface. Although beautiful to look at, frost can kill young plants.

Frozen windows

Condensation can also take place on other cold surfaces surrounded by warm air. Windows and car windscreens often become steamed up, and at freezing point they too become covered in frost. The ice crystals often join together to form beautiful feathery patterns.

Some animals are especially noticeable during cold weather, among them the owl:

'When icicles hang by the wall
And Dick the shepherd blows his nail,
And Tom bears logs into the hall
And milk comes frozen home in pail,
When blood is nipped and ways be foul,
Then nightly sings the staring owl...'

Love's Labour's Lost
by William Shakespeare (1564-1616)

FORECASTING SNOW

In the hands of the gods?

When the American explorer Robert Peary returned from the Arctic in 1909, Ootah, an Inuit guide on his expedition, told him: "The devil is asleep. . . or we should never have come back so easily."

Before people understood weather science, they believed that it was gods, spirits and demons who created ice and snow. The Inuit people believed in a spirit called Agloolik, who lived under the ice, and in Nootaikok, the spirit of icebergs. People have always prayed to their gods for an end to the cold weather. A big freeze could put a stop to fishing or hunting. A harsh winter could mean that food ran low and people died of cold. In the Middle Ages, heavy snows could cut a village off from the outside world for months at a time and snow could trap new-born lambs on the mountainside. Spring frosts could destroy seedlings and ruin crops. The end of winter was a cause for joy and was celebrated with festivals.

Searching for signs

People whose livelihood depended on fair weather, such as sailors, farmers and shepherds, searched for signs of foul weather and cold winters. They

▼ Heavy downfalls of snow can trap people if they are unexpected. Forecasting enables people to predict when snow is likely to fall.

▲ Farmers rely on forecasts to make sure they take their animals down to lower fields before the snows come.

An old wives' tale?

★ In the USA it was believed that if a groundhog came out of its burrow on 2 February and saw its own shadow, then it would scuttle back underground and the winter snows would continue for another six weeks.

★ In Britain an old rhyme makes exactly the same point: 'If Candlemas be fair and clear, there'll be two winters in the year.' Candlemas is also celebrated on 2 February. There is no proof that if that day is sunny, another bout of wintry weather will return.

watched the skies and learned which clouds and which winds brought snow. They studied plant and animal life for tell-tale clues. It was said that if there were a lot of berries in the autumn, or if leaves withered on the branch instead of falling quickly, the winter would be a very cold one.

Many of the old sayings are still known today. Some of them make sense, but many of them don't. Today we can rely on science instead to tell us whether icy conditions lie ahead.

◀ Many weather stations are in remote areas. Scientists visit the stations to collect information.

▼ Satellites photograph clouds and can follow the progress of storms. They send the information back to Earth.

WEATHER SCIENCE
Studying the weather
The scientific study of the weather is called meteorology. Scientists use many instruments to record the conditions. The maximum-minimum thermometer is a U-shaped tube which records the highest and lowest temperatures in a given period. A snow-gauge collects falling snow and measures its depth. An anemometer measures wind-speed and a barometer measures the pressure of the air on the Earth's surface.

Collecting information
All this information is collected at weather stations around the world. These may be based on land or on ships at sea. Planes and weather balloons are sent up to record weather conditions at high altitudes. Radar is used to track snow storms. A ring of weather satellites has been launched into space to scan the cloud systems on the Earth below.

Ice and snow in the clouds

Several kinds of clouds can contain ice or snow:

Cumulonimbus are huge, tall clouds, producing hail and heavy showers.

Cirrus form so high that it is freezing cold. They are made up of ice crystals.

Stratocumulus are common, flat-shaped clouds with dark parts. In winter they sometimes produce light snowfall.

Stratus are very low clouds. They rarely produce much snow.

Nimbostratus are large clouds that stretch over big areas of land and build up quite high. These are the clouds that cover the countryside with a thick, white carpet of snow during the winter.

▶ **Clouds are usually grouped according to the height, or altitude, at which they occur.**

Cirrus

Cumulonimbus

Stratocumulus

Stratus/nimbostratus

A WINTER WEATHER DIARY

1 Record the weather conditions over the winter months in a special notebook.

2 Check the temperature at the same time each morning and evening. Work out the average temperature at these times for each month.

3 Record whether there is a frost each morning.

4 Mark down any days on which there is snow, sleet or hail.

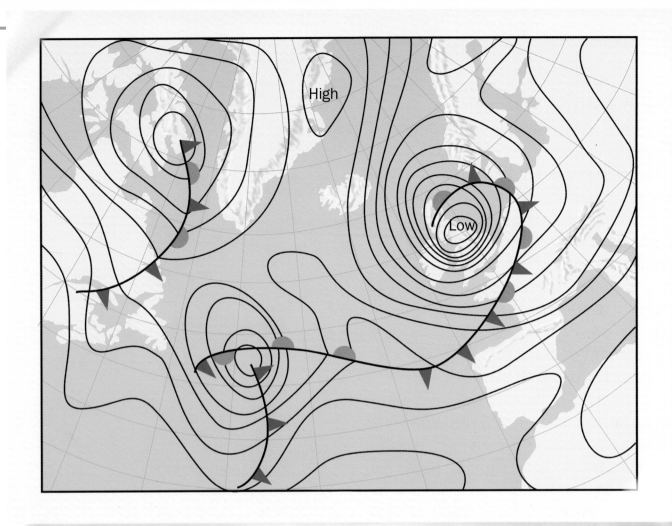

High

Low

Occluded (mixed) fronts   Cold fronts   Warm fronts   Isobars

GETTING THE PICTURE
Piecing together the information

The vast amount of information received from satellites and from thousands of weather stations and ships around the world is put together and studied with computers, so that we can gain a worldwide picture of the weather. Maps of weather systems can then be prepared, with lines showing places of equal temperature (isotherms) and equal air pressure (isobars).

▲ **The lines are called isobars. They join areas of equal pressure on a weather chart.**

Lines of warmer or cooler air are shown as warm or cold fronts. Simplified maps appear on television weather forecasts, with symbols such as black clouds, snowflakes and hail-stones. Wind directions may be shown with arrows and figures indicate likely temperatures.

Bad weather warnings

Shipping forecasts are broadcast on the radio. They include warnings of storms, poor visibility and ice. Forecasts of wind, snow, freezing fog and ice are also vital to aeroplane pilots, who receive the latest meteorological data before taking off from an airport or landing. Most airports have their own weather stations.

Advising mountaineers

Special reports are often issued to warn of weather conditions in the mountains. Climbers and walkers depend on forecasts of snow and ice for their safety. This is very important, as the weather at the base of the mountains when people set out may be very different from the conditions at the summit. Winter sport resorts also issue reports on snow conditions for skiing, and warn if rising temperatures and loose snow make avalanches likely.

First forecasts

★ The first weather reports to be published in a newspaper appeared in Britain in 1692.

★ In the USA, the first national weather service started in 1871. It was run by the US Army Signal Corps.

★ The first weather forecasts on radio were broadcast on the radio station 9XM in the USA in January 1921.

▼ **Weather forecasts give vital information about conditions, making it possible for people to plan their outdoor activities safely.**

◀ **Signs like this warn that avalanches can happen in the area.**

LIVING WITH THE COLD

Hunting for survival

The first people to live in the Arctic probably followed the herds of caribou. These animals were a valuable source of food and hunting was a way of life. As well as hunting caribou, seals and wolverines provided the meat and skins that were essential to living in such cold places. No parts of these animals were wasted – even the guts were scraped clean and thin and used to make windows. Today, the Inuit have domesticated the caribou and also use their milk to drink and make cheese.

Modern living

Today, most Inuit do not have to hunt to survive. The seas teem with fish and fish processing factories employ many local people. Oil has also been found deep underground, and other people earn their livings working to bring the oil to the surface.

Cold and dark

★ The coldest town in the world is Oymyakon in Siberia, Russia, which recorded a low of –68°C in 1933.

★ The long darkness of the Arctic winter means big electricity bills in northern cities. Helsinki, the capital of Finland, has 51 days of constant darkness in winter – but then it does have 73 days on non-stop daylight in summer.

▶ An Inuit hunter leaps between ice blocks in the Arctic.

◀ In Iceland, people can enjoy bathing outside even though the temperature is freezing cold. Warm water from underground geysers fills some lakes.

WILDLIFE IN WINTER

Keeping warm

Animals which live in cold climates have had to adapt to survive. Polar bears have coats of thick fur to keep warm and feet with pads designed for gripping slippery ice floes. Caribou have broad hooves designed for walking over the frozen bogs of the tundra. Musk-oxen have a dense furry undercoat to keep them warm in winter, but they have to moult to keep cool in summer.

Changing to the colour of ice

Many Arctic animals have white feathers or fur to camouflage them against the snow and ice. Some animals, such as Arctic hares and stoats, change their colour in autumn when the first snows arrive. Some animals go into hibernation in winter months. They become sleepy and their bodies slow down, so that they need less food. They spend the winter in cosy dens or nests.

Caribou

fallow deer

polar bear

brown bear

WALKING ON THE WILD SIDE

In the snow, animals footprints can be seen clearly. Different animals have differently shaped feet and, if you know what to look for, you can tell which animals have been there.

Watch animals walking over the snow and study their footprints. See the different shapes they leave behind. Deer, such as caribou, have split hooves which leave a distinctive mark. Dogs cannot retract or pull back their claws, which means the claws also leave a mark in the snow. Water birds, such as penguins, may have webbed feet or distinct toes if they do not live in water.

▲ **Penguins in Antarctica can survive colder weather than any other animal.**

Wildlife in abundance

The cold Arctic ocean is rich in marine life, attracting seals, whales, walruses, seabirds and fish. The pools of the Arctic tundra are the breeding grounds for swarms of insects and in summer many birds migrate north to feed on them. Other birds and mammals feed on the mosses, lichens and grasses of the tundra. No mammals live in the windy interior of Antarctica, but the coasts are the breeding grounds of many birds, including various kinds of penguin. Their plump bodies and sleek coats keep them warm as they fish in the icy waters.

Animal records

★ The record for the longest lie-in in winter is held by the Arctic ground squirrel of Canada and Alaska. It hibernates for nine months of the year.

★ The Arctic tern migrates each year between its Arctic breeding grounds and Antarctica, on the opposite side of the world. One bird was found to have travelled 22,530 kilometres.

PLANT LIFE
Surviving the cold

Plants need water, warmth and sunlight to grow well, but even so many of them have adapted to survive in the coldest Arctic frosts or on the highest, most exposed mountain peaks. Lichens form crusty or leafy patches on rock. Flowering plants in cold places often hug the ground or grow in crevices where they are sheltered from the bitter winds. Trees and shrubs which do grow higher may be bent and twisted by the wind. Only the tough needles and cones of pine and fir trees stand up well to harsh, snowy climates.

Growing in the cold

★ Flowering plants have been found growing in the bitterly cold Himalaya mountains at a height of 6,400 metres above sea level.

★ Sphagnum moss, found in northern temperate regions, has hollow cells in its stems and leaves. These can hold up to 20 times the moss's weight in water.

★ The custom of bringing fir trees into the house at Christmas and covering them with candles started in Germany, and spread from there in the 1800s. Today Christmas trees are popular even in hot countries where it never snows.

◀ **This region of Siberia is forested mainly with coniferous trees.**

Protecting tender shoots

Many of the plants we grow as food crops are put at risk by periods of cold weather. Frosts can kill off tender seedlings and shoots, which really need to be protected under glass. In some places bonfires or special stoves may be lit in fields to prevent frost forming on the ground overnight. In Iceland, the country's hot springs are used to heat greenhouses, so allowing vegetables and flowers to be grown in a frosty climate.

Wrapping plants up

Field crops may also be protected from the cold with plastic sheeting or with layers of straw. Strangely enough, a heavy snowfall may have a similar effect. It can act as a warm blanket, protecting the spring shoots below from chill winds and frosts. However the weight of the snow may also break branches and batter down plants.

Soil thaws in the summer

Permafrost

◀ **In winter, in the Arctic and Antarctic, all the moisture in the soil freezes. Only a few centimetres at the top thaws in the summer. The rest is permanently frozen and is called 'permafrost'.**

▲ **In Iceland, the cold weather would kill unprotected plants. Crops are grown in greenhouses heated by hot water from underground geysers.**

COLD PEOPLE

Humans are not as well protected against the cold and wind as many wild animals. But humans are very cunning, and in prehistoric times our ancestors learned to make fires and to stitch skins and furs together to keep warm.

Adapting to the cold

The human body does have several ways of keeping itself warm. In cold temperatures, we shiver. This jerking movement makes our muscles work harder and this keeps the body warm. To prevent heat loss from the skin, blood may stop flowing through the blood vesels near the surface. This may give the skin a bluish tinge when it is very cold.

Problems of the cold

The dangers of extreme cold are very serious indeed. A fall in body temperature to danger level is called hypothermia. It threatens the lives of the very young and the elderly each winter. It is also a risk to climbers if they are stranded on a mountain overnight. Chilling can be prevented by special 'survival blankets' which trap the body's heat.

◄ **In extremely cold conditions, frost can form wherever there is water vapour, even on people's eyebrows, eyelashes and clothing.**

★ When it is cold, our skin may rise up in 'goose bumps' or 'goose pimples'. This raises the little hairs on our body, which trap warm air next to our skin.

★ In cold weather you can see your breath. The water vapour in your breath condenses on meeting the cold air and forms a mini-cloud.

★ Wind increases the chill felt by the human body. When the temperature is 0°C, a wind speed of 40 kilometres per hour will make it feel like -18°C.

Icy noses, fingers and toes

Frostbite is a risk to people in very cold lands, and is a major problem for polar explorers and mountaineers. The skin and flesh freeze so that the blood won't flow freely. The nose, toes and fingers may change colour and lose all feeling.

◀ **Inuit living in the Arctic wear seal skins to protect against the cold. They always cover their heads as more heat is lost from the head than any other part of the body.**

LIVING WITH ICE
Peoples of ice and snow

Many different peoples have made their homes in the bitterly cold lands of the Arctic. No less than 26 different peoples, officially known as 'northern minorities', live in Siberia and along the Arctic shores of Russia.

The Saami people live in Lapland, the northernmost part of Scandinavia and Russia. Canada and Alaska are home to various Native American peoples and to the Inuit peoples, who also live in Russia and Greenland.

New ways of life

All the Arctic peoples have faced great changes this century and many are struggling to keep their way of life. Traditionally they have lived by hunting, fishing and whaling or by herding reindeer. Today however, whaling is strictly limited by international law, and the migrating routes of reindeer and caribou are blocked by new roads and towns. Many Saami can no longer earn their living from herding reindeer, so more and more of them are becoming farmers or fishermen.

Finding the past

The large countries to the south have brought in new industries, but often these offer little to local people whose

◀ **Many people including the Saami living in the cold lands are nomadic. This means they are often on the move, following their herds to greener pastures.**

▲ **Traditional ways of life still continue in some parts of the world. These Nenets women live in northern Siberia.**

languages and cultures are threatened by the newcomers. Many of the Arctic peoples now demand a greater say in the way they are governed and a revival of interest in their own culture. In 1999, the Canadian Inuit will be granted a large, self-governing territory which will be known as Nunavut.

Changing times

Mountain peoples around the world traditionally lived by leading their herds to high pastures during the summer, and bringing them back to the safety of farms in the valley before the winter snows come. Today many mountain villages have become busy resorts for winter sports fans, this can help the local economy, but some mountain forests and scrublands are being destroyed to make ski runs.

SNOW HOUSES

Today, many Arctic villages are made up of centrally heated modern houses and wooden cabins. They are often built around fish processing plants, mines or trading stores. Houses built in tundra regions risk collapse, for the endless freezing and thawing makes the soil slip and slide. However, many buildings are purpose-built to withstand extreme conditions. The scientific bases of the Antarctic, and the oil industry bases in northern Alaska offer warmth and comfort while the winds howl outside.

Building for snowy weather

Traditional housing about the Arctic circle once included tents made of reindeer or caribou skins, huts made out of stone and turf and the overnight hunting lodge of the Inuit, sometimes called an igloo. This small dome, made from angled blocks of frozen snow, is surprisingly warm and safe, and can withstand a blizzard that would blow down any tent.

▶ In the Inuit language, the word 'igloo' means 'dwelling'.

Frosty pipes

In lands which have cold winters, pipes and houses must be well wrapped up or insulated against the cold. Water expands when it freezes. When water pipes freeze, the expanding ice can crack them wide open. As the ice melts, water may pour out of the broken pipes and damage houses, bringing down ceilings and flooding rooms. Icy weather can cause all sorts damage in cities, as hailstones smash windows, snow brings down power lines and thaws bring floods.

▶ **Pedestrians trudge through Times Square in New York on 8 January 1996. A raging blizzard hit the city, leaving up to 45cm of snow. Airports and schools were closed and there was no mail delivery.**

GETTING MOVING

If ice and snow can damage the natural landscape, they can also make life hard in towns and cities. A harsh winter soon cracks a road, and houses built on the tundra risk collapsing as the soil becomes waterlogged or deep-frozen.

Sliding along

For fast transport across snow or ice, skis or runners may be fitted to vehicles. Traditionally sleds and sleighs were pulled over the ice by teams of dogs or horses, but today snowmobiles are commonly used in the Arctic. These are motorised sleds. Light aircraft too may be fitted with skis instead of wheels, so that they can land on the ice. Normal ships cannot pass through pack ice, but specially built, immensely strong icebreakers can smash a channel through the ice to Arctic ports such as Arkhangelsk in Russia.

◄ When the snow lies deep on the ground, snowblowers can clear a path for traffic.

▼ Ships with specially reinforced hulls can smash through the frozen seas. They are called icebreakers.

◀ **Roads have to be kept free from snow and ice to stop street cars and other road users from skidding.**

Travelling over ice

When roads are covered in snow and ice, it is hard for wheels to grip the surface. Cars and lorries may skid and lose control. Roads and airport runways often have to be cleared by snow-ploughs. Grit may be spread from lorries to provide friction or gripping power. This can also be gained by fitting sets of chains over the tyres of a vehicle. Antifreeze must be put in a car's radiator to keep it working in cold winters. Trains can only run in cold countries if the points on the track are electrically heated so that they do not ice up.

High flyers

★ 'Icing up' is a common problem in flying. In 1919 John Alcock and Arthur Whitten Brown made the first non-stop crossing of the Atlantic Ocean by air. During the 16-hour flight Brown had to climb out of the cockpit to knock lumps of ice off the plane, so that he could read the petrol gauge.

SALT VERSUS ICE

1 A sprinkling of salt makes it harder for the road surface to freeze up. Test this for yourself.

2 Fill an ice tray with tap water and place it in the freezer box of the fridge.

3 At the same time, fill another ice tray with water mixed with salt and place that in the freezer box too. Which tray freezes the first?

WINTER SPORTS

Long ago, travellers in snowy countries learned to strap runners of wood or bone to their feet in order to slide over ice or snow. These were the first skis and skates. Snowshoes, large frames shaped like tennis rackets, were also laced to the feet. These spread the weight of the body, making it easier to cross soft snow. People have always enjoyed slipping and sliding for fun, and so skiing and skating soon became popular sports.

Skiing contests include downhill racing, ski-jumping and slalom races in which skiers zigzag around markers in the snow. Skaters may race at speed or dance to music on specially prepared ice rinks. Rinks are also used for the fast team sport of ice-hockey. Winter sports fans also make use of bobsleighs, toboggans, snowboards and ice-yachts, which sail over frozen lakes on steel runners.

▲ **Snowboarding is an increasingly popular winter sport.**

LAYERS FOR WARMTH

In cold weather you put on extra layers of thicker clothes for warmth. The clothes trap a layer of air close to your body. Your body warms the air, which makes it less dense. The air cannot escape from among the fibres into the cooler surrounding air, and cold air cannot enter to replace it. This effect is called insulation.

Warm clothing is essential for survival in the snow. Clothes made out of animal skins are best and can protect against temperatures as low as -60°C.

Cotton T shirt

hat

warm jacket

trousers

wool jumper

Lightweight Underwear

boots

wool socks

Snow rescue

Many people visit mountain resorts around the world to enjoy winter sports and climbing. But if an accident happens, rescue teams may have to be sent out, either on foot or using helicopters to reach remote mountain peaks. Tracker dogs such as German shepherds and St Bernards sniff for people buried in avalanches.

Skis and Skates

★ A ski found in Sweden is thought to be about 4,500 years old.

★ The world's longest ice skating race is the 'Eleven Towns Tour', first held in the Netherlands in the 1600s. It is run over 200 kilometres of rivers and canals and attracts thousands of competitors.

◀ In Scandinavian countries, reindeer racing is a popular sport.

THE FUTURE

▼ During the last Ice Age, about one million years ago, deep sheets of ice covered about one-third of all the land on the planet.

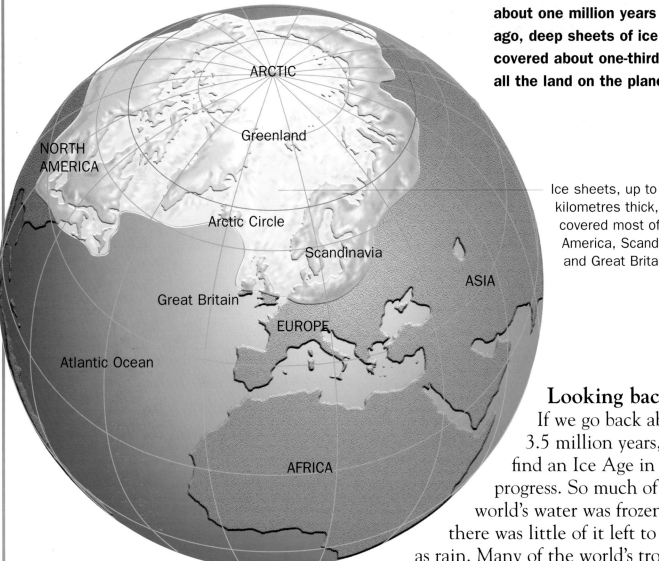

ARCTIC

Greenland

NORTH AMERICA

Arctic Circle

Scandinavia

ASIA

Great Britain

EUROPE

Atlantic Ocean

AFRICA

Ice sheets, up to 3 kilometres thick, covered most of North America, Scandinavia, and Great Britain.

Looking back

If we go back about 3.5 million years, we find an Ice Age in progress. So much of the world's water was frozen that there was little of it left to fall as rain. Many of the world's tropical forests were replaced by dusty grasslands. If we move forward to about 10,000 years ago, the climate was becoming warmer and so much polar ice melted that large areas of the world were flooded by rising sea levels. Today, ice sheets are only found at the North and South Poles.

Past and future

The world's climate does not remain the same. Very, very slowly, over many millions of years, our planet experiences warmer or colder periods. These changes are still happening today.